Revision *notes*

Intermediate 2
Physics
revision notes

✕ Neil Short ✕

Published by
Leckie & Leckie
8 Whitehill Terrace
St Andrews
Fife
KY16 8RN
T: 01334 475656
F: 01334 477392
E: enquiries@leckieandleckie.co.uk
W: www.leckieandleckie.co.uk

Special thanks to
BRW (creative packaging), Caleb Rutherford (cover design), Pumpkin House (concept design and illustration), Tony Wayte (copy-editing and proofreading) David Pearce (content review).

ISBN 1-84372-150-3

A CIP Catalogue record for this book is available from the British Library.

Printed by The Thanet Press

® Leckie & Leckie is a registered trademark

Leckie & Leckie Ltd is a division of Granada Learning Limited, part of ITV plc.

Contents

Introduction

The Intermediate 2 Physics course contains four units:

1 **Mechanics and Heat**
2 **Electricity and Electronics**
3 **Waves and Optics**
4 **Radioactivity**

Unit 1 and unit 2 each make up a third of the course. The final third of the course is made up of units 3 and 4 which are both shorter.

In order to gain a full qualification in Intermediate 2 Physics you have to achieve three different targets:

1 Complete **Unit Tests** in each of the four units.
   In the two larger unit tests (*Mechanics and Heat and Electricity and Electronics*) you will probably have to score about 24 out of 40 to pass. In the two smaller units (*Waves and Optics and Radioactivity*) the tests are shorter and a mark of about 12 out of 20 should be sufficient to put a smile on your face!
   Check with your teacher about the pass marks before you sit each test.
   If you don't reach the required standard in a test you are able to try again in the future.

2 Perform and write up an **Experiment** to a satisfactory standard.
   The experiment can be taken from any of the four units.

3 Pass the final **External Examination**.

Depending on your performance you can achieve an A, B or C pass.

For extra revision and invaluable practice, it is worth keeping an eye out for copies of the official SQA Past Papers or Leckie and Leckie's *Questions in Intermediate 2 Physics*.

Action Verbs

When teachers and examiners assess your written work, they focus on both your Knowledge and Understanding (**KU**) and your Problem Solving (**PS**) skills. This book makes use of five "Action Verbs" to help you improve these skills.

To sharpen up your **Knowledge and Understanding** look for the **Memorise**, **State**, **Understand** and **Describe** symbols.

The keys to success in **Problem Solving** are the **Memorise** and **Solve** symbols.

Memorise

Most Physics exams will eventually provide you with a list of formulas (like **F=ma**) but it is still a very good idea to know your formulas well in advance of getting to the exam. Do the hard work early and learn each formula when you meet it.

State

You can be asked to state basic facts like: "*one newton is a force which gives a mass of one kilogram an acceleration of one metre per second per second*". Use these Revision Notes to learn the basic facts. It is also useful to draw up your own list of these basic facts as it will help you to remember them.

Understand

Physics becomes much easier when you really understand the important ideas. If at first you have trouble understanding an idea, do not give up. Try talking about it with a classmate or make time to see your teacher.

Describe

You may be asked in an exam to describe an experiment or you may be asked to describe a process (like nuclear fission). These Revision Notes contain partly completed descriptions. Complete them by filling in the blanks. The correct answers are supplied.

Solve

You will be faced with lots of Physics problems in the exam so practice in problem solving is essential. Try the problems in these Notes. Hide the answers then check them when you have finished.

Chapter 1

Mechanics and heat

SPEED, DISTANCE, TIME

Memorise

Average speed = $\dfrac{\text{total distance travelled}}{\text{time}}$ $\bar{v} = \dfrac{s}{t}$

The **unit of speed** (or velocity) is the **metre per second m/s**.

Solve

A van leaves Edinburgh at 9:00am and makes a **return** journey to Galashiels (53km away), followed by a single 218 km journey to Aberdeen, arriving at 3:00pm. Find the average speed:

ANSWER
$\bar{v} = \dfrac{s}{t}$ $\dfrac{(53 + 53 + 218)}{6 \times 60 \times 60} \times 1000 = 15$ m/s

Memorise

Instantaneous speed = $\dfrac{\text{distance}}{\text{time}}$ $v = \dfrac{s}{t}$

Understand

Average speed and instantaneous speed both come from the formula $v = \dfrac{s}{t}$ so what's the difference?

Average speed is usually measured over a longer time. Instantaneous speed is measured over a very small time interval. It is the speed at an instant. Ideally it should be measured over a time interval that is nearly zero.

Solve

A picture of a bullet taken by a high speed camera shows it has moved 5 cm in one ten-thousandth of a second. Find the instantaneous speed in both **a** cm/s and **b** m/s.

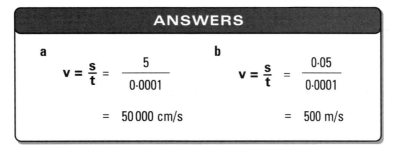

ANSWERS

a

$$v = \frac{s}{t} = \frac{5}{0 \cdot 0001}$$

$$= 50\,000 \text{ cm/s}$$

b

$$v = \frac{s}{t} = \frac{0 \cdot 05}{0 \cdot 0001}$$

$$= 500 \text{ m/s}$$

Describe

An experiment to measure average speed for the journey from X to Y.

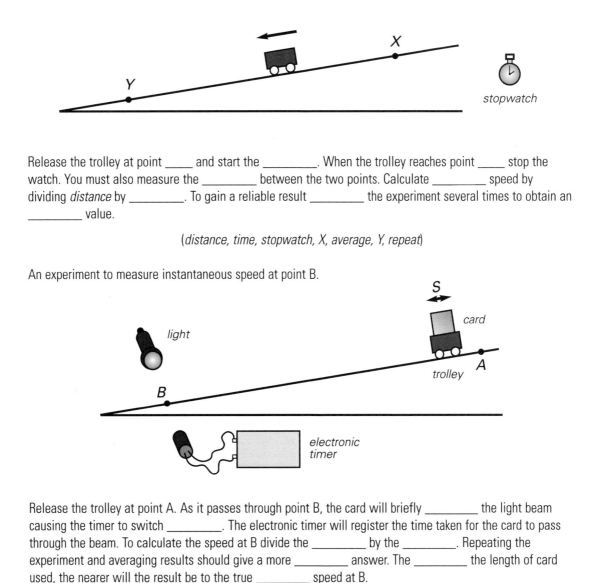

stopwatch

Release the trolley at point _____ and start the _____. When the trolley reaches point _____ stop the watch. You must also measure the _____ between the two points. Calculate _____ speed by dividing *distance* by _____. To gain a reliable result _____ the experiment several times to obtain an _____ value.

(*distance, time, stopwatch, X, average, Y, repeat*)

An experiment to measure instantaneous speed at point B.

S

card

light

trolley

A

B

electronic
timer

Release the trolley at point A. As it passes through point B, the card will briefly _____ the light beam causing the timer to switch _____. The electronic timer will register the time taken for the card to pass through the beam. To calculate the speed at B divide the _____ by the _____. Repeating the experiment and averaging results should give a more _____ answer. The _____ the length of card used, the nearer will the result be to the true _____ speed at B.

(*reliable, shorter, card-length, time, instantaneous, break, on*)

SCALARS AND VECTORS

Understand

The quantities you meet in physics: speed, mass, force, etc. are either scalars or vectors. **Scalars** have a **size only**. **Vectors** have a **size** and they point in a particular **direction**.

Memorise

Distance, time and speed are scalars.

Displacement and velocity are vectors.

$$\text{Velocity} = \frac{\text{displacement}}{\text{time}} \qquad v = \frac{s}{t}$$

Understand

Look at the following worked example to help you understand the difference between scalars and vectors. A dog walks 400 m East to a lamp post, stops briefly then goes 300 m North to join a waiting friend. This takes 5 minutes.

Find

a the distance travelled

b the displacement

c the average speed in m/s

d the average velocity in m/s

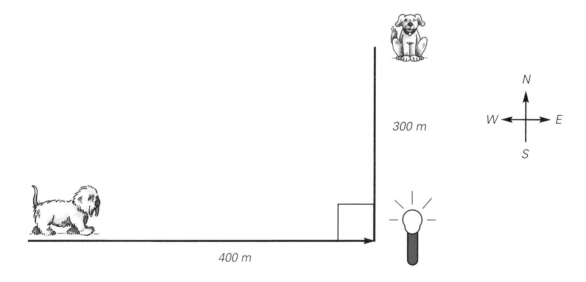

ANSWERS

a distance = 400+300 = 700m

b

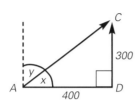

displacement can be found either by scale drawing or by using the following maths: (Pythagoras Theorem):

displacement = AC

$AC^2 = 400^2 + 300^2$ *so* AC = 500 m

$$\tan \text{ angle } x = \frac{\text{opposite}}{\text{adjacent}} = \frac{300}{400} = 0.75$$

so x = 37°

We often give the direction as a **bearing**. This is the angle from North. In the diagram the bearing is shown as angle y. The final answer to the question is:

displacement is 500m at a bearing of 053°.

c average speed $= \dfrac{\text{distance}}{\text{time}} = \dfrac{700}{5 \times 60} = 2.3$ m/s

d average velocity $= \dfrac{\text{displacement}}{\text{time}} = \dfrac{500}{5 \times 60} = 1.7$ m/s bearing 053°

Memorise

$$\text{Acceleration} = \frac{\text{change of velocity}}{\text{time}} = \frac{(\text{final velocity} - \text{initial velocity})}{\text{time}} \qquad a = \frac{v - u}{t}$$

The **units of acceleration** are **metres per second squared m/s²**.

Acceleration is a vector.

The equation for acceleration above can be rearranged to obtain   **v = u + at**

Solve

1 A car accelerates from rest and reaches 30 m/s in 10 seconds. Find the acceleration.

2 A train travelling at 50 m/s applies the brakes and reduces its velocity to 10 m/s in 80 seconds. Find the acceleration.

3 A plane travelling at 80 m/s accelerates at 4 m/s² for a time of one minute.
Find its final velocity.

ANSWERS

1 $a = \dfrac{v - u}{t} = \dfrac{30 - 0}{10} = 3 \text{ m/s}^2$

2 $a = \dfrac{v - u}{t} = \dfrac{10 - 50}{80} = -\dfrac{40}{80} = -0.5 \text{ m/s}^2$

3 $v = u + at$
   $= 80 + 4 \times 60$
   $= 80 + 240$
   so $v = 320 \text{ m/s}$

Understand

Velocity/Time graphs

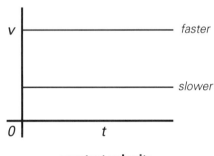

constant velocity

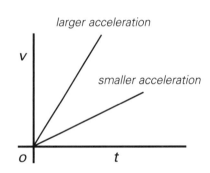

uniform acceleration

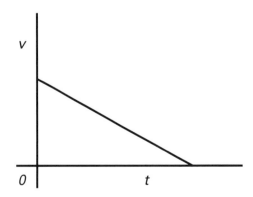

uniform deceleration (negative acceleration)

Acceleration can be found from the **gradient** of a velocity/time graph.

Solve

Find the accelerations from these velocity/time graphs:

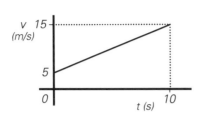

a

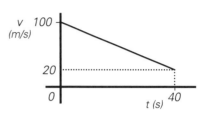

b

ANSWERS

a $a = \dfrac{v - u}{t} = \dfrac{15 - 5}{10} = \dfrac{10}{10} = 1 \text{ m/s}^2$

b $a = \dfrac{v - u}{t} = \dfrac{20 - 100}{40} = \dfrac{-80}{40} = -2 \text{ m/s}^2$

Understand

The distance travelled can be found from the area under a speed/time graph:

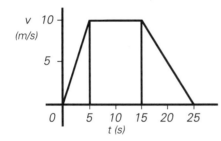

 10 + 10 + 10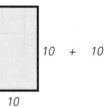

Area = 25 + 100 + 50 = 175   Distance = 175 m

Solve

A cyclist accelerates from rest to 5 m/s in 10 seconds. She continues at that speed for 20 seconds then brakes to a halt in a further 15 seconds. Sketch a speed/time graph of her motion and use it to find:

a   her acceleration during the first 10 seconds

b   the total distance travelled.

ANSWERS

a $a = \dfrac{v - u}{t} = \dfrac{5 - 0}{10} = 0.5 \text{ m/s}^2$

b   Area = 162·5
       so distance = 162·5 m

FORCES

Understand

Forces can change the **shape**, **speed** or **direction** of travel of an object.

State

Force is a **vector**

The **unit of force** is the **newton, N**

A force of 1 N will give a mass of 1 kg an acceleration of 1 m/s^2

Understand

Force can be measured using a Newton balance (spring balance). The spring stretches in proportion to the force applied. If a 3 N force stretches a spring by 2 cm then a 6 N force will stretch it by 4 cm.

Solve

The diagrams show a spring being stretched with different forces. Find the missing values, X, Y and Z.

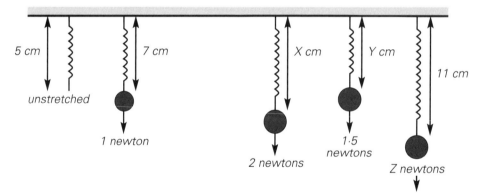

ANSWER

1 N stretches the spring from 5 cm to 7 cm length

1 N stretches the spring by 2 cm

2 N will stretch it by 4 cm

so X = 5 + 4 = 9 cm

1·5 N will stretch the spring by 3 cm

so Y = 5 + 3 = 8 cm

The unknown force has stretched the spring by 6 cm

It will require 3 N to stretch it by 6 cm

so Z = 3 N

State

Friction is a force which can oppose the motion of an object.

Understand

screech

Cases where friction forces are of benefit.   Cases where it is important to keep friction to a minimum.

State

Weight is the force of gravity acting on any mass. The **unit of weight** is the **newton, N**.

Understand

Mass, m is a measure of the **amount of matter** in an object. It depends on the number of atoms in the object and the size of these atoms. The **unit of mass is the kg**. The mass of an object would not change, even if it was taken away from Earth to another planet.

Weight, W is a gravity dependent quantity. The weight of an object depends on

•   the mass
•   the gravitational field strength

Gravitational Field Strength, **g**, is a measure of the strength of gravity on a particular planet. It tells you the weight of a 1 kg mass on that planet i.e. the force of gravity pulling down on a kilogram. The **unit of gravitational field strength** is the **newton per kilogram, N/kg**.

Memorise

Weight = mass x gravitational field strength, **W = mg**.

Solve

Use the values of gravitational field strengths to find the answers to the questions:

Earth	Moon	Jupiter
10 N/kg	1·6 N/kg	26 N/kg

1   Find the weights of these objects on **a** the Earth **b** the Moon

 (i) 60 kg   *(ii)* 1 tonne   *(iii)* 30 g

2 A lunar rover vehicle weighs 320 N on the Moon. What is its weight on Earth?

3 The stone and the brass mass shown below are perfectly balanced on Earth.

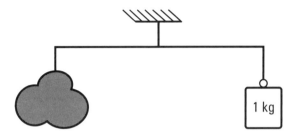

Which of these brass masses would need to be used to balance the stone on Jupiter?

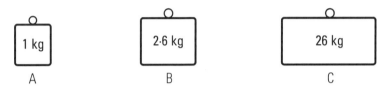

1 kg	2·6 kg	26 kg
A	B	C

ANSWERS

1 **a** (i) 600 N   (ii) 10 000 N   (iii) 0·3 N

     b (i) 96 N   (ii) 1600 N   (iii) 0·048 N

2 Weight on the Moon W = mg *so* 320 = m × 1·6

     so mass, m = 20 kg

     so Weight on Earth mg = 200 × 10 = 2000 N

3 On Earth, the stone balances the 1 kg therefore the stone has a mass of 1 kg. The stone's mass does not change on Jupiter, so A is the correct answer.

Understand

When forces are balanced, the forces cancel out to produce a resultant force of zero. Here are three examples of balanced forces:

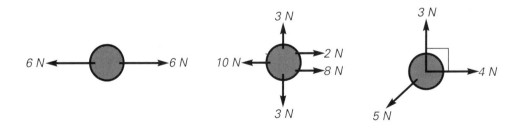

Newton's First Law of Motion

If the forces on a stationery object are balanced, it will remain at rest. If the forces on a moving object are balanced, it will continue moving at a constant speed in a straight line.

This law tells us that if forces on an object are balanced, then its speed won't change and the direction in which its moving won't change.

Solve

These vehicles have all reached their top speeds. What are the values of the resistive forces R acting on them?

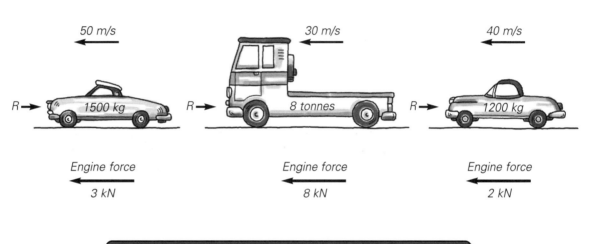

50 m/s	30 m/s	40 m/s
R→ 1500 kg	R→ 8 tonnes	R→ 1200 kg
Engine force	Engine force	Engine force
3 kN	8 kN	2 kN

ANSWERS		
a 3 kN	b 8 kN	c 2 kN

(If you really understand Newton's 1st Law you will realise that the values of mass and speed have no effect on the answers.)

Understand

Newton's Second Law of Motion

If an unbalanced force acts on an object, it will accelerate. The acceleration will vary directly will the force and inversely the mass. The law can be written in the form of an equation:

Unbalanced Force = mass x acceleration

The unbalanced force can also be called the resultant force.

Memorise

F = ma

Solve

1 Find the resultant forces:

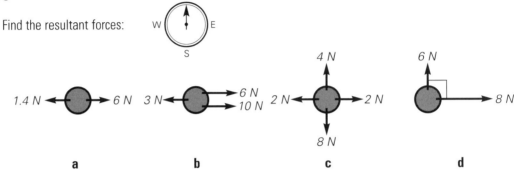

2 Find the accelerations:

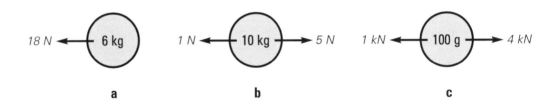

3 Find the acceleration of this rocket:

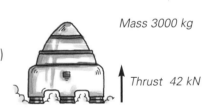

Mass 3000 kg

a On take-off from the Moon (no air resistance; g = 1·6 N/kg)

b On take-off from Earth (air resistance = 3 kN; g = 10 N/kg)

c In deep space

Thrust 42 kN

ANSWERS

1 **a** 4·6 N East   **b** 13 N East   **c** 4 N South

d $R^2 = 6^2 + 8^2 = 100$
so R = 10 N
angle a = 37° (bearing = 053°)

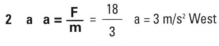
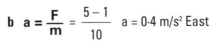

2 **a** $a = \dfrac{F}{m} = \dfrac{18}{3}$ a = 3 m/s² West

b $a = \dfrac{F}{m} = \dfrac{5-1}{10}$ a = 0·4 m/s² East

c $a = \dfrac{F}{m} = \dfrac{4000-1000}{0·1}$ a = 30 000 m/s² East

ANSWERS (continued)

3 Draw a free body diagram to show the forces:

a   ↑ *Thrust 42 000 N*

●

↓ *Weight mg = (3000 x 1·6) = 4800 N*

$$a = \frac{F}{m} = \frac{42\,000 - 4\,800}{3\,000} = \frac{37\,200}{3\,000} \quad a = 12\cdot4 \text{ m/s}^2$$

b   ↑ *Thrust 42 000 N*

Air resistance   ↓↓   *Weight mg = (3000 x 10)*
3000 N   ↓↓   *= 30 000 N*

●

$$a = \frac{F}{m} = \frac{42\,000 - 33\,000}{3\,000} = \frac{9\,000}{3\,000} \quad a = 3 \text{ m/s}^2$$

c   ↑ *42 000 N*

●

$$a = \frac{F}{m} = \frac{42\,000}{3\,000} = a = 14 \text{ m/s}^2$$

Understand

If we drop an object, it will accelerate downwards under the influence of the force of gravity (weight). We call this acceleration *the acceleration due to gravity*.

Using the equation $a = \frac{F}{m}$ we get:

$$\text{acceleration due to gravity} = \frac{\text{Force of gravity}}{\text{mass of object}}$$

$$a = \frac{mg}{m}$$

so $a = g$

Acceleration due to gravity is equivalent to gravitational field strength.

State

The motion of a projectile can be split into two independent parts: constant horizontal velocity and vertical acceleration due to gravity.

Understand

If an object is travelling horizontally and there are no friction forces, it will go on at a constant speed (Newton's 1st Law).

If an object is allowed to fall from rest and there are no friction forces, it will accelerate downwards at 10 m/s².

If a projectile has been fired horizontally it will do both of the above things at the same time. This will cause it to travel in a curved path called a parabola.

So, projectile motion consists of two independent parts: constant horizontal speed and constant vertical acceleration.

Look carefully at these worked examples:

1   ⬤ ➔ *5 m/s*   This ball has been fired horizontally at 5 m/s along a completely smooth surface.

     a Draw a speed/time graph for the first 4 s of its journey.

     b What is its speed after 2 s?

     c How far has it travelled after 3 s?

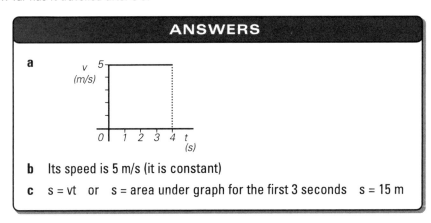

ANSWERS

a

b Its speed is 5 m/s (it is constant)

c s = vt   or   s = area under graph for the first 3 seconds   s = 15 m

2   This ball falls from rest.

     a Draw a speed/time graph for the first 4 s of its journey.

     b What is its speed after 2 s?

     c How far has it travelled after 3 s?

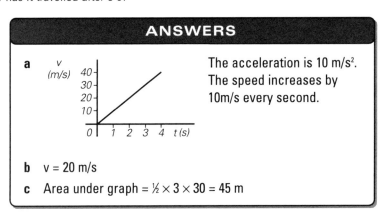

ANSWERS

a   The acceleration is 10 m/s². The speed increases by 10m/s every second.

b   v = 20 m/s

c   Area under graph = ½ × 3 × 30 = 45 m

3 This ball has been projected horizontally off the cliff at 5 m/s. It hits the ground 3 s later at Point X.

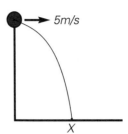

    a Draw   (i) a horizontal speed time graph

                (ii) a vertical speed time graph of its motion

    b How far from the foot of the cliff is Point X?

    c How high is the cliff?

ANSWERS

a

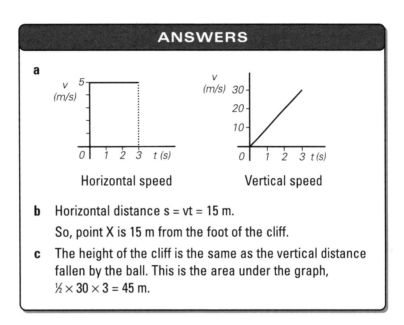

Horizontal speed   Vertical speed

b Horizontal distance s = vt = 15 m.

    So, point X is 15 m from the foot of the cliff.

c The height of the cliff is the same as the vertical distance fallen by the ball. This is the area under the graph, $\frac{1}{2} \times 30 \times 3 = 45$ m.

Understand

Newton's Third Law of Motion

If object A exerts a force on object B, then object B will exert an equal but opposite force on object A. Here are some examples of the law:

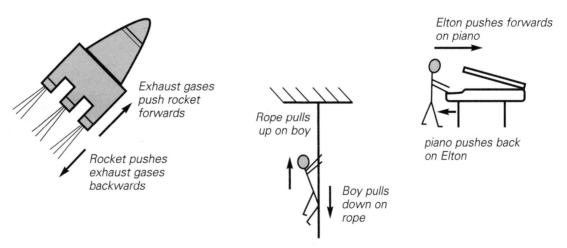

Exhaust gases push rocket forwards

Rocket pushes exhaust gases backwards

Rope pulls up on boy

Boy pulls down on rope

Elton pushes forwards on piano

piano pushes back on Elton

The forces are sometimes called Newton Pairs.

Solve

There are two sets of Newton Pairs in the example of a book on a table.
Can you identify both sets?

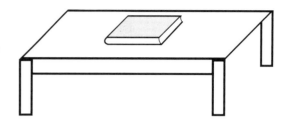

ANSWERS

1 Downwards force of book on table/upwards force of table on book.

2 Weight of book (i.e. downwards force of Earth on book)/upwards force of book on Earth.

MOMENTUM

State

Momentum = mass x velocity   **p = mv**

The **units of momentum are kg m/s.**

Momentum is a vector quantity.

The **Law of Conservation of Momentum** states that when objects collide the total momentum remains unchanged (provided no external forces interfere).

Total momentum before collision = Total momentum after collision.

Solve

A 3 kg trolley travelling at 5 m/s collides with a stationary 2 kg trolley. The trolleys stick together and move off at speed v.

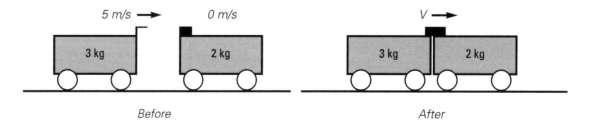

Before		*After*

Find: **a**   the total momentum before the collision

      b   the total momentum after the collision

      c   the speed v

ANSWERS

a Total momentum before collision = (3 x 5) + (2 x 0) = 15 + 0 = 15 kg m/s

b Total momentum after collision is the same, 15 kg m/s

c mv = 15 *so* 5v = 15 *so* v = 3 m/s

ENERGY

Understand

Energy exists in different forms, heat, kinetic, potential, etc. Energy can change from one form to another (when work is done) but the **Law of Conservation of Energy** says that energy cannot be created from nothing and it cannot be destroyed.

State

Work is a measure of the amount of energy transferred from one form to another.

Memorise

Work = Force x distance $E_W = Fs$ (or W = Fs)

The unit of work is the joule, J.

Solve

How much work is done in moving the blocks from A to B in each case:

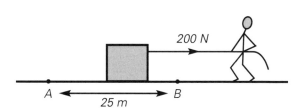
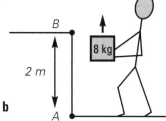

ANSWERS

a $E_W = Fs = 200 \times 25 = 5000$ J

b $E_W = Fs$
The minimum force to raise the block at a steady speed will be:
mg = $8 \times 10 = 80$ N.
so $E_W = 80 \times 2 = 160$ J

Understand

When work is done against the force of gravity to raise an object upwards, the object gains **Gravitational Potential Energy, E_p**.

Memorise

$E_p = mgh$

The unit of energy is the joule, J.

Solve

How much gravitational potential energy is gained by each object in moving from X to Y?

a

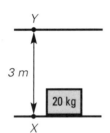

b

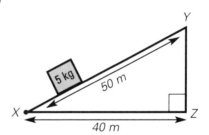

ANSWERS

a $E_p = mgh = 20 \times 10 \times 3 = 600$ J

b $E_p = mgh$. To find h we must work out height YZ.
Use Pythagoras Theorem. You should find that YZ is 30 m.

      so $E_p = 5 \times 10 \times 30 = 1500$ J.

Memorise

Kinetic energy = 1/2 × mass × (speed)²   **$E_k = \frac{1}{2} mv^2$**

Solve

Find the kinetic energies:

a

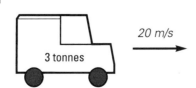

20 m/s

3 tonnes

b

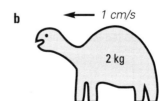

1 cm/s

2 kg

ANSWERS

a $E_k = \frac{1}{2} mv^2 = \frac{1}{2} \times 3000 \times (20)^2 = \frac{1}{2} \times 3000 \times 400 = 60\,000\,J$

b $E_k = \frac{1}{2} mv^2 = \frac{1}{2} \times 2 \times (0.01)^2 = 0.0001\,J$

Memorise

$$Power = \frac{energy}{time} \qquad Power = \frac{work\ done}{time}$$

The unit of power is the watt (W).

One watt is the same as **one joule per second**.

Solve

Find the power output in each case:

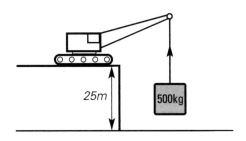

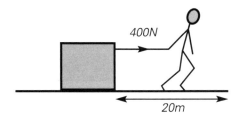

a The crane lifts the load in 20 s

b John pulls the block 20 m in 40 s.

ANSWERS

a $P = \dfrac{E_p}{t} = \dfrac{mgh}{t} = \dfrac{500 \times 10 \times 25}{20} = 6250\ W$

b $P = \dfrac{E_w}{t} = \dfrac{Fs}{t} = \dfrac{400 \times 20}{40} = 200\ W$

Understand

When a machine transforms energy from one form to another, some energy is wasted as heat (due to friction forces). No machine is completely (100%) efficient.

Memorise

$$\% \text{ efficiency} = \frac{\text{useful energy output}}{\text{energy input}} \times 100$$

$$\% \text{ efficiency} = \frac{\text{useful power output}}{\text{power input}} \times 100$$

Solve

1   The crane in the last question has a motor of power rating 10 kW. Find its percentage efficiency when lifting the load.

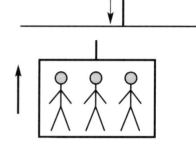

2   The power rating of the lift motor is 5000 W. The efficiency is 80%. Find the useful power output.

3   The lift can travel up 100 m in 3 min 20 s. How large a weight can the motor raise?

ANSWERS

1 $\% \text{ efficiency} = \dfrac{\text{power out}}{\text{power in}} = \dfrac{6250}{10\,000} \times 100 = 62.5\%$

2 $\text{efficiency} = \dfrac{\text{power out}}{\text{power in}} = \dfrac{80}{100} = \dfrac{P}{5000}$

$P = \dfrac{5000 \times 80}{100} = 4000 \text{ w}$

3 $P = \dfrac{E}{t} = \dfrac{Fs}{t}$

so $4000 = \dfrac{F \times 100}{200}$

so $F = 8000 \text{ N}$

The lift motor raises a weight of 8000 N (a mass of 800kg)

HEAT

Understand

To get equal masses of different materials to go up in temperature by one Celsius degree, you need different amounts of heat.

The diagrams show how much energy is needed to raise the temperature of three different materials by one Celsius degree.

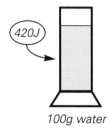

100g water

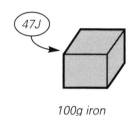

100g iron

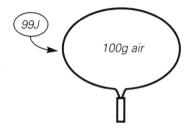

100g air

Memorise

The specific heat capacity, c of a substance is the **energy required to raise the temperature of 1kg by 1°C**.

$$c = \frac{E_h}{m\Delta T} \qquad E_h = cm\Delta T$$

The **units of specific heat capacity are J/kg°C**

(The term ΔT is called *delta T*. It means the difference in temperature. Ask your teacher about ΔT)

Solve

1 A pupil heats a beaker containing 100 g of water for 5 minutes using a 20 W electric heater. The temperature rises from 18°C to 30°C. Find:

 a the energy output from the heater

 b the pupils' value for the specific heat capacity (s.h.c.) of water.

2 Lead has a s.h.c. of 130 J/kg°C. Make an estimate of the temperature rise in a lump of lead which falls to the ground from a height of 42 m.

ANSWERS

1 a $E = Pt = 20 \times 5 \times 60 = 6000$ J

 b $c = \dfrac{Eh}{M\Delta T} = \dfrac{6000}{0.1 \times 12}$

 $= 5000$ J/kg °C

2 If the lead falls, it loses gravitational potential energy. If we assume that this energy changes to heat and that all of it goes to raising the temperature of the lead then:

 $mgh = cm\Delta T$

 so $gh = c\Delta T$ *so* $\Delta T = \dfrac{gh}{c} = \dfrac{10 \times 42}{130}$

 so $\Delta T = 3°C$

(In practice, heat would be lost elsewhere and ΔT would be less.)

Understand

For a substance to change from solid to liquid or from liquid to gas it must gain latent heat. For a substance to change from gas to liquid or from liquid to solid it must lose latent heat.

Example:

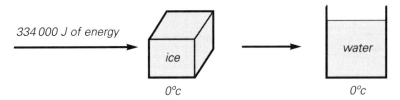

For 1 kg of ice at 0°C to change into 1kg of water at 0°C it is necessary to add 334 000 J of heat.

While a substance is changing state (solid to liquid) or (liquid to gas) there is no change in temperature.

Memorise

Specific Latent Heat, L of a substance is the **energy required to change the state of 1 kg** of the substance at its melting point or boiling point.

The **units of specific latent heat are J/kg**.

Understand

Every substance has two values of L:

       specific latent heat of fusion (solid/liquid)

       specific latent heat of vaporisation (liquid/gas)

Chapter 2

Electricity and electronics

ELECTROSTATICS

State

Conductors contain free electrons. These electrons can move freely through the conductor.

Metals are **good conductors**.

Insulators don't have free electrons. They don't conduct electricity.

Most **non-metals** are **insulators**.

Describe

An electric current is a drift of _____ particles (usually electrons) through a _____ . The size of the current depends on the amount of _____ passing a point every _____ .

(second, charge, conductor, charged)

Memorise

Current = $\dfrac{\text{charge}}{\text{time}}$ $I = \dfrac{Q}{t}$

The **unit of charge** is the **coulomb, C**.

The **unit of time** is the **second, s**.

The **unit of current** is the **ampere, A**.

When **one coulomb** of charge passes a point in **one second** the current is **one ampere**.

Solve

1 Find the currents when 30 coulombs of charge pass a point in:

  a 3 seconds

  b 1 minute

2 How much charge passes through the resistor:

  a in 30 seconds

  b in one hour

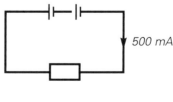

500 mA

3 How long will it take for 60 C of charge to pass through the bulb?

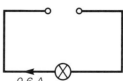

0·6 A

ANSWERS

1 a $\quad I = \dfrac{Q}{t} = \dfrac{30}{3} = 10\ A$

b $\quad I = \dfrac{Q}{t} = \dfrac{30}{60} = 0.5\ A$

2 a $\quad I = \dfrac{Q}{t}\quad so\quad Q = It = 0.5 \times 30 = 15\ C$

b $\quad Q = 0.5 \times 60 \times 60 = 1800\ C$

3 $\quad I = \dfrac{Q}{t}\quad so\quad t = \dfrac{Q}{I} = \dfrac{60}{0.6} = 100\ s$

CIRCUITS – SERIES AND PARALLEL

Understand

Electric current is a tricky idea. Try to imagine an **electric current** as similar to the **flow of water** through a pipe.

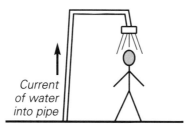

Power shower with 150 cm³ of water passing per second.

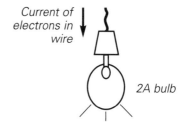

Light bulb with 2 C of charge passing per second.

Memorise

Circuit symbols

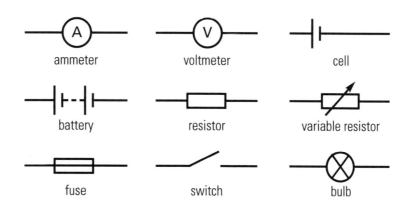

State

Voltage is a measure of the number of joules of **energy** given to each coulomb of **charge** in a circuit.

The **unit of voltage** is the **volt**, **V**.

Understand

In an electric circuit, the battery gives energy to the charges (electrons), which pass it on to the bulbs, resistors and other components.

The **voltage** across a component is sometimes called the **potential difference, pd**. Charges give up some of their energy as they pass through a component like a resistor. They leave with less energy than they entered it. We say that the potential has dropped.

This drop in potential is the **potential difference**. It is measured in **volts** using a voltmeter.

How to connect ammeters correctly in a circuit.

Ammeters are connected in **series** (because they have to measure the amount of charge passing through a component every second).

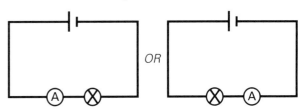

Voltmeters are connected in **parallel** (because they have to sense the difference in the potentials before and after charges move through the component).

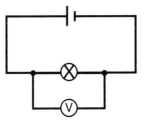

State

In a **series** circuit the **current** is the **same in all components**.

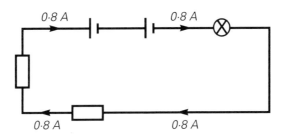

Understand

The current has to be the same around a series circuit because current is a measure of the number of electrons passing and there are no escape routes. Electrons may lose energy as they pass through each component but they don't disappear so the number of electrons does not change.

 State

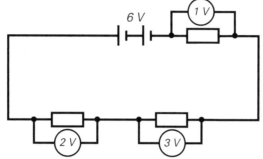

In a **series** circuit the **sum** of the **pds across the components equals the supply voltage**. (1 + 3 + 2 = 6)

 Understand

The battery voltage is a measure of the energy supplied to the charges. In a series circuit, the battery has to supply some energy to each resistor so it divides the total voltage (6 V) among the resistors. (Big resistors demand more energy than small ones so they grab a bigger part of the battery voltage.)

 State

In a **parallel** circuit the **sum** of the **currents in the branches equals the supply current**. (0·8 + 1·2 + 0·5 = 2·5)

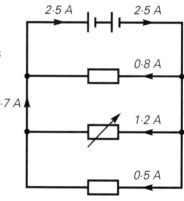

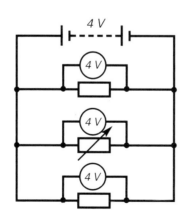

In a **parallel** circuit the **voltage** across each branch is the **same**.

 Understand

It is tempting to think of three voltages, each one equal to four volts.

In fact, there is only **one** voltage of four volts in the parallel circuit.

Memorise

	SERIES	PARALLEL
CURRENT	same	add
VOLTAGE	add	same

Solve

Use the information given in the circuits below to find all the meter readings.

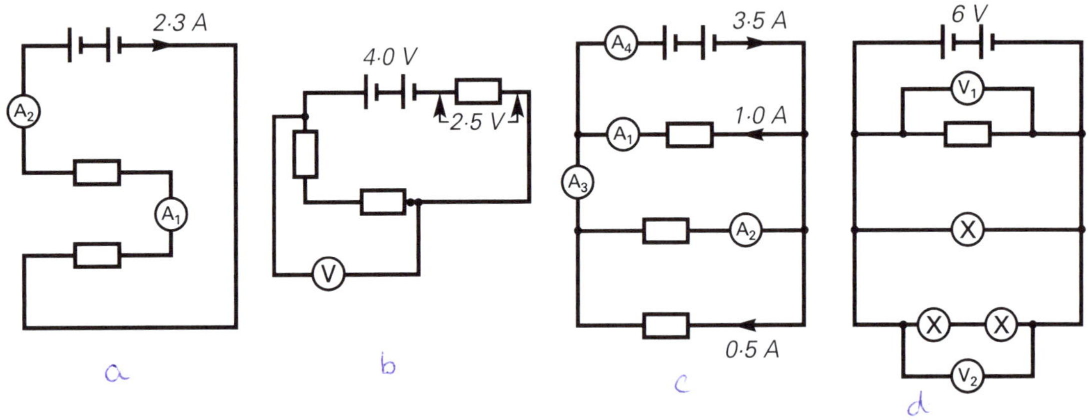

ANSWERS

a A_1 = 2·3 A A_2 = 2·3 A

b V = 1·5 V

c A_1 = 1·0 A A_2 = 2·0 A A_3 = 2·5 A A_4 = 3·5 A

d V_1 = 6 V V_2 = 6 V

Understand

Ohm's Law

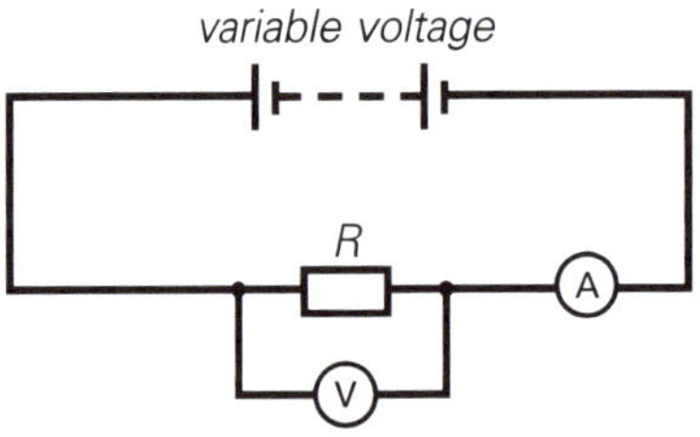

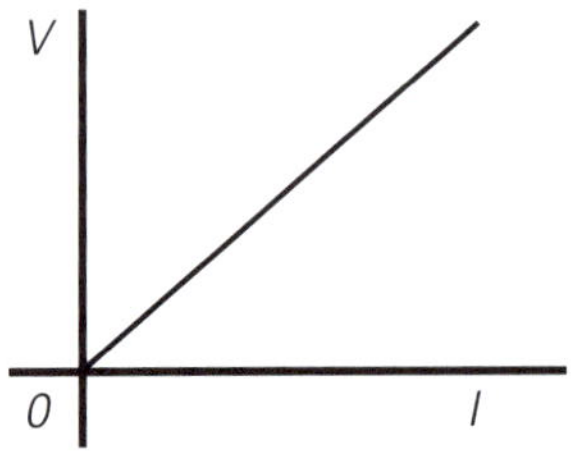

The graph of V against I is a straight line through the origin. This means that V varies directly with I for a resistor (at constant temperature).

Memorise

Another way of stating Ohm's Law is: $\dfrac{V}{I} = R$

R is a constant called resistance.

The **unit of resistance** is the **ohm, Ω**.

The formula is often written as **V = I R**.

Solve

Find the missing values, V, I and R:-

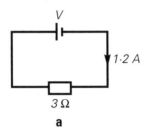

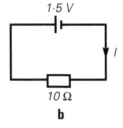

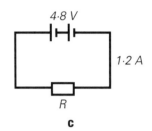

a	b	c

ANSWERS

a V = I R = 1·2 × 3 *so* V = 3·6 V

b $I = \dfrac{V}{R} = \dfrac{1·5}{10}$ *so* I = 0·15 A

c $R = \dfrac{V}{R} = \dfrac{4·8}{1·2}$ *so* R = 4 Ω

Now check out these harder questions:

All the cells in the circuits below have a voltage of 2V. Find the meter reading and the value of resistor R.

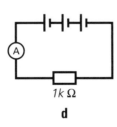

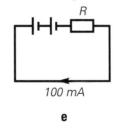

d	e

ANSWERS

d $I = \dfrac{V}{R} = \dfrac{3 \times 2}{1\,000} = \dfrac{6}{1\,000}$ *so* I = 0·006 A (6 mA)

e $R = \dfrac{V}{I} = \dfrac{2 \times 2}{0·1} = \dfrac{4}{0·1}$ *so* R = 40 Ω

Memorise

Resistors in **series**

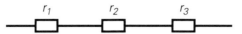

Total resistance $R_T = r_1 + r_2 + r_3$

Resistors in **parallel**

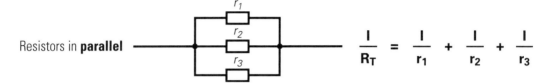

$$\frac{1}{R_T} = \frac{1}{r_1} + \frac{1}{r_2} + \frac{1}{r_3}$$

Solve

Find the total resistance R_T in each case:

a

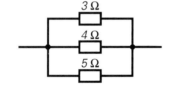

b

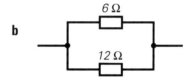

c

d

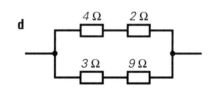

ANSWERS

a $R_T = 3 + 4 + 5 = 12\ \Omega$

b $\dfrac{1}{R_T} = \dfrac{1}{6} + \dfrac{1}{12} = \dfrac{3}{12}$ *so* $R_T = \dfrac{12}{3} = 4\ \Omega$

c $\dfrac{1}{R_T} = \dfrac{1}{3} + \dfrac{1}{4} + \dfrac{1}{5} = \dfrac{47}{60}$ *so* $R_T = \dfrac{60}{47} = 1.28\ \Omega$

d $\dfrac{1}{R_T} = \dfrac{1}{4+2} + \dfrac{1}{3+9} = \dfrac{1}{6} + \dfrac{1}{12}$ *so* $R_T = 4\ \Omega$
(same as b)

Memorise

The total resistance of a series circuit must always be greater than the resistance of the largest resistor.

The total resistance of a parallel circuit must always be less than the resistance of the smallest resistor.

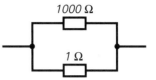

If you are in doubt about the parallel resistance statement, try this:

You should find that R_T is $\dfrac{1000}{1001}$ which is less than 1 Ω.

Understand

Techniques for solving circuit problems.

Series example

Find the current in the 4 Ω resistor and the voltages across the 2 Ω and 6 Ω resistors.

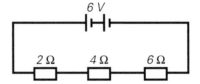

- The current is the **same** in each resistor. (There is only one current.)

- Use the formula $I = \dfrac{V}{R}$, Current $= \dfrac{\text{Total voltage}}{\text{Total resistance}} = \dfrac{6}{12} = 0.5\,A$

- The current in the 4 Ω resistor is **0·5A**

Now use the formula V = IR

    Voltage across 2 Ω resistor   V = IR = 0·5 x 2 = **1V**

    Voltage across the 6 Ω resistor   V = IR = 0·5 x 6 = **3V**

Parallel example

Find the voltage across the 4 Ω and 12 Ω resistors and the reading on the ammeter.

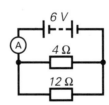

- The voltage across each resistor is the same in this circuit

- The voltage across the 4 Ω resistor is 6V. The voltage across the 12 Ω is also 6V

- The ammeter reads the total (supply) current. Use the formula $I = \dfrac{V}{R}$

    Supply current $= \dfrac{\text{Total voltage}}{\text{Total resistance}} = \dfrac{6}{R_T}$

- Find R_T from $\quad \dfrac{1}{R_T} = \dfrac{1}{4} + \dfrac{1}{12} \quad$ so $R_T = \mathbf{3\,\Omega}$

- $I = \dfrac{6}{R_T} = \dfrac{6}{3} = \mathbf{2A}$

- Try an alternative way of finding supply current. Find the currents in the 4 Ω and 12 Ω resistors separately then add them together. Get help if necessary.

Solve

In both circuits below, find the current in the 12 Ω resistor and the voltage across the 3 Ω resistor.

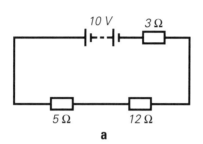

a

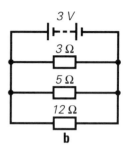

b

ANSWERS

a $I = \dfrac{V}{R} = \dfrac{10}{20} = 0.5\,A$

$V = IR = 0.5 \times 3 = 1.5\,V$

b $I = \dfrac{V}{R} = \dfrac{3}{12} = 0.25\,A$

$V = 3\,V$

Understand

The Potential Divider Circuit

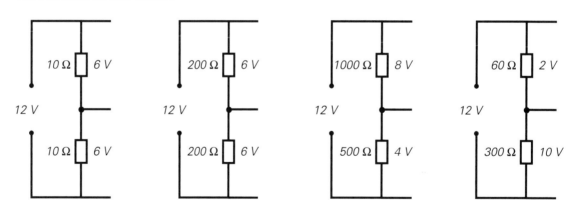

The supply voltage is divided up across the two resistors in proportion to their resistance values.

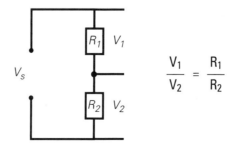

$$\frac{V_1}{V_2} = \frac{R_1}{R_2}$$

Memorise

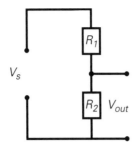

$$V \text{ out} = \frac{R_2}{(R_1 + R_2)} \times V_s$$

Solve

Find the output voltage readings on the voltmeters.

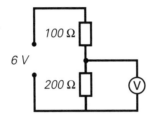

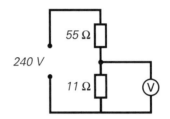

ANSWERS

a $V \text{ out} = \dfrac{200}{(100 + 200)} \times 6 = \dfrac{200}{300} \times 6 = \mathbf{4V}$

b $V \text{ out} = \dfrac{11}{(55 + 11)} \times 240 = \dfrac{11}{66} \times 240 = \mathbf{40V}$

(Notice if both the resistances are in k Ω you don't need to write 11 000, 55 000 etc.)

CIRCUITS - POWER

State

When there is an electric current in a component energy is transformed.

For example, when current flows in a bulb, electrical energy is transformed to heat and light. A motor transforms electrical energy to kinetic energy and heat.

The **unit of electrical energy** (and all other forms of energy) is the **joule, J**.

Power is a measure of the **amount of energy transformed every second**.

Memorise

$$\text{Power} = \frac{\text{Energy}}{\text{Time}} \qquad P = \frac{E}{t}$$

The **unit of power** is the **watt, W**.

One watt is equivalent to **one joule per second**.

Solve

1 Find the power rating of a light bulb which transforms one kilojoule of energy every 10 seconds.

2 How long does it take a 3 kW electric fire to use up 60 000 J of energy?

3 How much energy is dissipated as heat in one hour by a 20 W immersion heater?

ANSWERS

1 $P = \dfrac{E}{t} = \dfrac{1000}{10} = 100 \text{ W}$

2 $t = \dfrac{E}{P} = \dfrac{60\,000}{3\,000} = 20 \text{ s}$

3 $E = Pt = 20 \times 60 \times 60 = 72\,000 \text{ J}$

Memorise

Power = Voltage × Current

P = VI

Solve

a A 230 V mains heater draws a current of 4·5 A when switched to a LOW setting. Find its power rating.

b When switched on at a HIGH setting the power rating is 2760 W. How much current is drawn on HIGH setting?

ANSWERS

a $P = VI = 230 \times 4{\cdot}5 = 1035 \text{ W}$

b $I = \dfrac{P}{V} = \dfrac{2760}{230} = 12 \text{ A}$

Understand

Two extra formulae for power can be worked out using a combination of V = IR and P = VI.

P = VI = (IR)I   *so* **P = I²R**

$P = VI = V\left(\dfrac{V}{R}\right)$ *so* $\mathbf{P = \dfrac{V^2}{R}}$

Solve

Find the power dissipated in each circuit below:

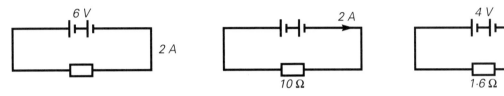

ANSWERS

a P = VI = 6 × 2 *so* P = 12 W

b P = I²R = 4 × 10 *so* P = 40 W

c $P = \dfrac{V^2}{R} = \dfrac{16}{1\cdot6}$ *so* P = 10 W

AC AND DC

Understand

The term **d.c.** stands for **direct current** – a drift of electrons always in the same direction. Direct current is produced by cells, batteries and d.c. power supplies.

d.c. circuit

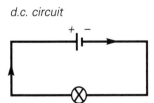

Electrons travel away from the negative terminal and round the circuit towards the positive terminal. The voltage always "pushes" in the same direction.

The term **a.c.** stands for **alternating current**. In a.c., electrons move for a time in one direction then they move back in the opposite direction under the influence of an a.c. voltage. The number of times this happens in one second is called the **frequency**. Alternating current is produced by a.c. power supplies.

State

In the UK the mains frequency is 50 Hz.

Understand

Two different values of a.c. voltage:

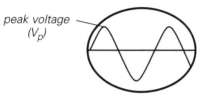

peak voltage (V$_p$)

Peak Voltage Vp

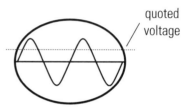

quoted voltage

Quoted voltage is approximately equal to $\dfrac{V_p}{1\cdot4}$

Because the peak voltage only occurs for an instant of time twice in each cycle it is a rather optimistic value. It is better to use the quoted (or effective) value. This will provide the same power as a d.c. voltage. In the diagrams below, bulbs A and C will light with the same brightness, but bulb B will be dimmer.

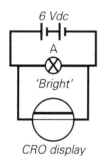

6 Vdc

A

'Bright'

CRO display

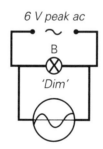

6 V peak ac

B

'Dim'

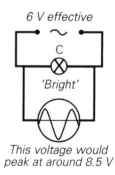

6 V effective

C

'Bright'

This voltage would peak at around 8.5 V

ELECTROMAGNETISM

State

A current carrying wire has a magnetic field around it.

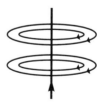

The field lines around a straight wire are circular.

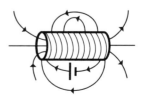

The field around a coil of wire is like that of a bar magnet.

Understand

You can show the difference between a strong magnet and a weak magnet like this:

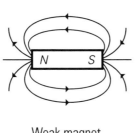

Weak magnet
Field lines far apart

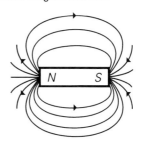

Strong magnet
Field lines closely packed together

A **voltage can be induced** across a conductor if it **moves** through **magnetic field lines**.

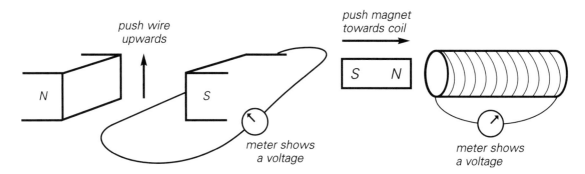

State

The **size** of the **voltage** depends on:

- the **strength of the magnetic field**
- the **speed of movement**
- the **angle** at which the conductor cuts the field lines
- the **number of turns** in the coil

Understand

You can sum up the first three points above by saying that the voltage depends on the number of magnetic lines cut every second.

In a coil, each loop of wire has a voltage induced across it and these individual voltages add up.

This explains why the number of turns in the coil affects the voltage.

Solve

Look at the values of induced voltages in the first four examples. Use the values to find voltages V_1, V_2, V_3 and V_4 in the second four examples. Magnets marked A are strong. Magnets marked B are weak.

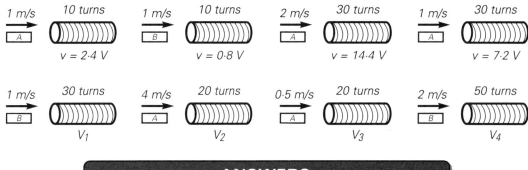

1 m/s $\rightarrow$ [A] 10 turns	v = 2·4 V
1 m/s $\rightarrow$ [B] 10 turns	v = 0·8 V
2 m/s $\rightarrow$ [A] 30 turns	v = 14·4 V
1 m/s $\rightarrow$ [A] 30 turns	v = 7·2 V

1 m/s $\rightarrow$ [B] 30 turns	V_1
4 m/s $\rightarrow$ [A] 20 turns	V_2
0·5 m/s $\rightarrow$ [A] 20 turns	V_3
2 m/s $\rightarrow$ [B] 50 turns	V_4

ANSWERS

$V_1 = 2·4$ V $V_2 = 19·2$ V $V_3 = 2·4$V $V_4 = 8·0$ V

TRANSFORMERS

State

A transformer can increase (step up) or decrease (step down) an alternating voltage.

Memorise

$$\frac{N_s}{N_p} = \frac{V_s}{V_p}$$

N – number of turns of wire

S – secondary

P – primary

$\dfrac{N_s}{N_p}$ is called the **turns ratio**

Solve

Use the equation above to find the missing quantities.

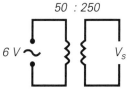

50 : 250

6 V ~ V_s

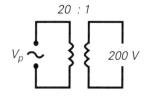

20 : 1

V_p ~ 200 V

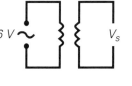

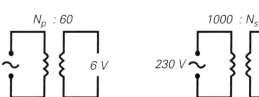

N_p : 60

2 V ~ 6 V

1000 : N_s

230 V ~ 115 V

ANSWERS

a $V_s = 30$ V

b $V_p = 4000$ V

c $N_p = 20$ turns

d $N_s = 500$ turns

Understand

Transformers **can increase voltage** but they **cannot increase energy** (or power). At the best, an ideal transformer would transform all the power in the primary circuit into power in the secondary circuit. No power would be lost.

In an ideal transformer Power in primary = Power in secondary

$$V_p I_p = V_s I_s \quad so \quad \frac{V_s}{V_p} = \frac{I_p}{I_s}$$

So we can write down a complete formula: $\dfrac{N_s}{N_p} = \dfrac{V_s}{V_p} = \dfrac{I_p}{I_s}$

Example

Find **a** V_s

   b I_s

   c I_p

   in this circuit

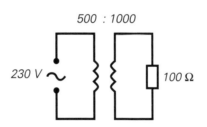

a $\dfrac{N_s}{N_p} = \dfrac{V_s}{V_p}$ *so* $\dfrac{1000}{500} = \dfrac{V_s}{230}$ *so* $500 V_s = 230 \times 1000$ *so* $V_s = 460$ V

b $I_s = \dfrac{V_s}{R} = \dfrac{460}{100} = 4.6$ A

c $\dfrac{N_s}{N_p} = \dfrac{I_p}{I_s}$ *so* $\dfrac{1000}{500} = \dfrac{I_p}{4.6}$ *so* $500 I_p = 4.6 \times 1000$ *so* $I_p = 9.2$ A

Understand

In the circuit above:

- the secondary voltage depends on primary voltage and turns ratio.
- the secondary current depends on secondary voltage and resistance.
- the primary current depends on the secondary current.

Solve

1 Find Vs, Is and Ip in these circuits:

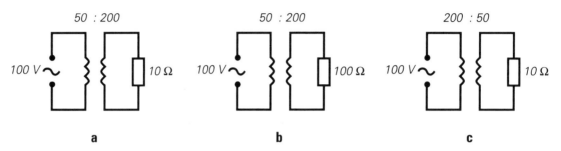

2 If the bulb is operating at normal brightness, calculate
   a the turns ratio
   b the secondary current
   c the primary current

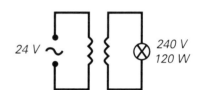

ANSWERS

1 a $\dfrac{N_s}{N_p} = \dfrac{V_s}{V_p}$ *so* $\dfrac{200}{50} = \dfrac{V_s}{100}$ *so* $50\,V_s = 200 \times 100$ *so* **$V_s = 400\ V$**

$I_s = \dfrac{V_s}{R} = \dfrac{400}{10} = \mathbf{40\ A}$

$\dfrac{N_s}{N_p} = \dfrac{I_p}{I_s}$ *so* $\dfrac{200}{50} = \dfrac{I_p}{40}$ *so* $50\,I_p = 40 \times 200$ *so* **$I_p = 160\ A$**

b $V_s = \mathbf{400\ V}$

$I_s = \dfrac{400}{100} = \mathbf{4A}$

$I_p = \mathbf{16\ A}$

c $\dfrac{N_s}{N_p} = \dfrac{V_s}{V_p}$ *so* $\dfrac{50}{200} = \dfrac{V_s}{100}$

so $200\,V_s = 50 \times 100$

so **$V_s = 25V$**

$I_s = \dfrac{25}{10} = \mathbf{2.5\ A}$

$\dfrac{N_s}{N_p} = \dfrac{I_p}{I_s}$ *so* $\dfrac{50}{200} = \dfrac{I_p}{2.5}$

so $200\,I_p = 50 \times 2.5$

so **$I_p = 0.625A$**

ANSWERS (continued)

2 a $\dfrac{N_s}{N_p} = \dfrac{V_s}{V_p}$ *so* $\dfrac{240}{24} = \dfrac{10}{1}$ **The ratio is 10:1**

b $P = VI$ *so* $I_s = \dfrac{P}{V_s} = \dfrac{120}{240} = 0.5\ A$

c $\dfrac{N_s}{N_p} = \dfrac{I_p}{I_s}$ *so* $\dfrac{10}{1} = \dfrac{I_p}{0.5}$ *so* $I_p = 5\ A$

ELECTRONICS – INPUT/OUTPUT DEVICES

Memorise

Some output devices and their energy transformations.

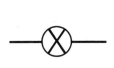

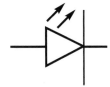

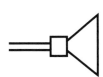

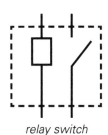

bulb
electrical ➜ light

light emitting diode (LED)
electrical ➜ light

loudspeaker
electrical ➜ sound

relay switch

Understand

An LED conducts electricity in one direction only.

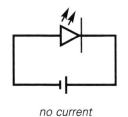

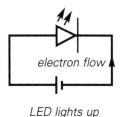

electron flow

no current

LED lights up

LEDs normally operate at voltages of about 2 volts. They often need a protective resistor in series if the power supply voltage is too big. You can calculate the value of a suitable resistor as follows:

V across LED should be 2 V

so V across resistor R should be 4 V (series voltages add)

I through LED should be 20 mA (0·02 A)

I through resistor R should also be 0·02 A

so $R = \dfrac{V}{I} = \dfrac{4}{0.02} = \mathbf{200\ \Omega}$

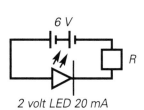

6 V

R

2 volt LED 20 mA

Solve

The LED in the circuit below is rated at 2 V, 10 mA. Find the size of the resistor required to let the LED operate normally.

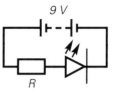

ANSWERS

$$V = 9 - 2 = 7\ V$$

$$R = \frac{V}{I} = \frac{7}{0\cdot01} = 700\ \Omega$$

Memorise

Some input devices

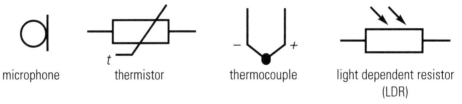

| microphone | thermistor | thermocouple | light dependent resistor (LDR) |

The resistance of the thermistor can decrease as temperature increases.

The resistance of the LDR decreases as light intensity increases.

Understand

Thermistors and LDRs can be used in potential dividers.

Example

The LDR has a resistance of 200 Ω in daylight and 4800 Ω in darkness. Find the readings on meters

a in daylight   **b** in darkness

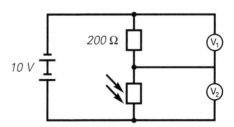

Daylight $V_1 = \dfrac{200}{200 + 200} \times 10 = 5\ V,\ V_2 = 5\ V$

Darkness $V_1 = \dfrac{200}{(200 + 4800)} \times 10 = 0\cdot4\ V,\ V_2 = 9\cdot6\ V$

Solve

A thermistor has a resistance of 1 k Ω at 100°C and 3 k Ω at 30°C. Find the voltmeter readings

a at 100°C **b** at 30°C

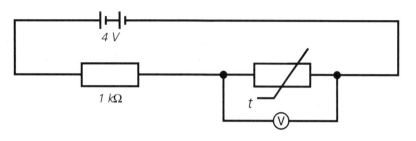

ANSWERS

a at 100°C $V = \dfrac{1}{1+1} \times 4 = \textbf{2V}$

b at 30°C $V = \dfrac{3}{1+3} \times 4 = \textbf{3V}$

TRANSISTOR CIRCUITS

Memorise

The symbols for:

An NPN transistor

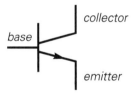

A MOSFET

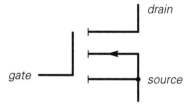

Understand

Both transistors and mosfets can act as electronic switches.

When the voltage across the base and emitter V reaches a certain positive value (usually about 0·7V) the transistor switches on. Current can then flow through the collector, base and emitter between X and Y allowing the bulb to light up.

When the voltage across the gate and source Vgs reaches a certain positive value the mosfet switches on. Current can then flow from the source Y through to the drain X allowing the bulb to light up.

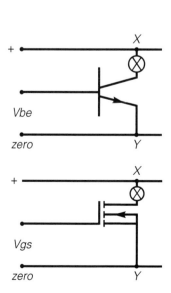

AMPLIFICATION

Memorise

Some devices which make use of amplifiers:

telephone, television, PA system, some hospital equipment

Understand

In audio amplifiers the signal increases in amplitude (but not frequency).

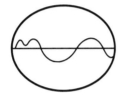 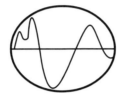

Memorise

Voltage gain $= \dfrac{\textbf{Output voltage}}{\textbf{Input voltage}}$ $\qquad$ $\text{gain} = \dfrac{V_{out}}{V_{in}}$

Solve

Find the missing values in these amplifier circuits:

$V_{in} = 50mV$ — [Gain = 100] — V_{out} $\qquad$ V_{in} — [Gain = 20] — $V_{out} = 0.5V$ $\qquad$ $V_{in} = 1mV$ — [Gain] — $V_{out} = 1.5V$

ANSWERS

a $\quad 100 = \dfrac{V_{out}}{50}$ so $V_{out} = 100 \times 50 = 5000 \text{ mV} = \textbf{5 V}$

b $\quad 20 = \dfrac{0.5}{V_{in}}$ so $V_{in} = \dfrac{0.5}{20} = \textbf{0.025V}$ (or 25 mV)

c $\quad \text{Gain} = \dfrac{1.5}{0.001} = \textbf{1500}$

Chapter 3

Waves and Optics

BASIC WAVE IDEAS

State

Waves transfer energy from one place to another. For example, infrared waves transfer heat from the Sun to the Earth. Water waves can transfer energy across oceans. Given enough time, the energy can smash large rocks down into grains of sand.

Memorise

$$\text{speed} = \frac{\text{distance}}{\text{time}} \qquad v = \frac{s}{t}$$

The **unit of speed** (or velocity) is the **metre per second, m/s**.

Describe

An experiment to measure the speed of sound in air.

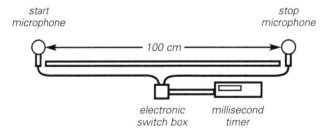

distance travelled by sound, d = 100 cm
time on timer, t = 0·003 s

Solve

A climber claps his hands and hears the echo from a distant cliff face 1·5 seconds later. If the speed of sound in air is 340 m/s how far is the climber from the cliff face?

ANSWERS

$$v = \frac{s}{t} \qquad s = vt$$

$$= 340 \times 1{\cdot}5 = \mathbf{510\ m}$$

So the distance the sound has travelled is 510 metres. The climber must be 255 metres from the cliff face. Think about it!

State

Radio and TV signals travel as electromagnetic waves. Like light waves, they travel at 300 million metres per second, 3×10^8 m/s.

Solve

Our Sun is approximately 150 million kilometres away from the Earth. How long does it take light to travel from the Sun to the Earth?

ANSWERS

$$v = \frac{s}{t} \quad so \quad t = \frac{s}{v} = \frac{150 \times 10^6 \times 10^3}{3 \times 10^8}$$

$$= \mathbf{500\ s} \ \ (\text{or } 8 \text{ min } 20 \text{ s})$$

Understand

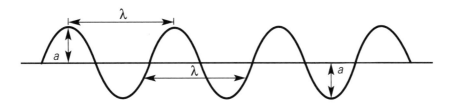

a Amplitude	a measure of the **energy** carried by the wave.
λ Wavelength	the **distance between any two corresponding points** on the wave (in **metres, m**).
f Frequency	the **number of waves per second** (in **hertz, Hz**).
T Period	the **time for one wave to pass a given point** (in **seconds, s**).

Solve

1 Which wave has:

   a the smallest amplitude

   b the longest wavelength

   c the lowest frequency

   d the shortest period?

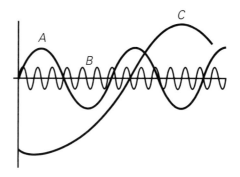

2 If the wave pattern below was produced in 0·5 s, find:

   a the amplitude

   b the wavelength

   c the frequency

   d the period

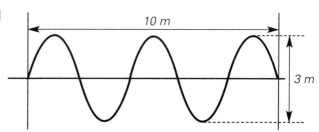

ANSWERS

1 **a** B   **b** C   **c** C   **d** B

2 **a** $\frac{1}{2} \times 3 = \mathbf{1·5\,m}$ **b** $2\frac{1}{2}\lambda = 10\,m$ *so* $\lambda = \mathbf{4\,m}$

    c $f = 2\frac{1}{2}$ waves in 0·5 s *so* 5 waves in 1 s *so* $\mathbf{f = 5\,Hz}$

    d 5 waves occur in 1 s *so* 1 wave occurs in 1/5 s $\mathbf{T = 0·2\,s}$

Understand

In a transverse wave, the particles of the material which the waves pass through vibrate at 90° to the wave direction.

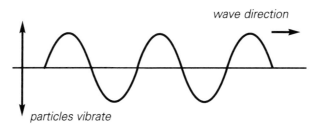

In a longitudinal wave, the particles vibrate backwards and forwards in the same direction as the wave.

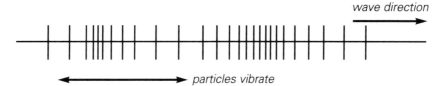

Sound waves are longitudinal. Water waves are transverse. All electromagnetic waves are transverse.

Memorise

The wave equation   speed = frequency x wavelength

$$v = f\,\lambda$$

Solve

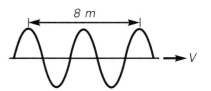

1   This wave has a frequency of 5 Hz. Find the wave speed, v.
2   If blue light has a wavelength of 4 x 10⁻⁷ m, find its frequency.

ANSWERS

1   $v = f\lambda = 5 \times 4 = $ **20 m/s**

2   $f = \dfrac{v}{\lambda} = \dfrac{3 \times 10^{8}}{4 \times 10^{-7}}$

   $= 7 \cdot 5 \times 10^{14}$ Hz.

State

The order of waves in the electromagnetic spectrum starting with the highest frequency

Gamma rays, x-rays, ultraviolet, visible, infrared, microwaves, TV, radio

Highest frequency   Lowest frequency

REFLECTION

Understand

The terms **angle of incidence i, angle of reflection r, normal, N**

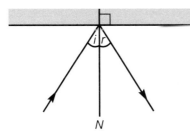   

Memorise

The law of reflection   **Angle of incidence = angle of reflection**
                              i = r

Understand

The principle of reversibility of light. If the direction of a ray of light is reversed, it still takes exactly the same path.

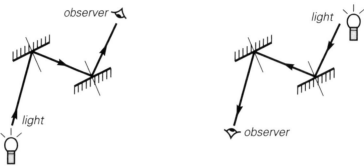

The effect of curved reflectors

Receiver

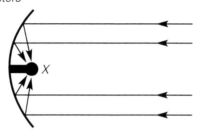

All the energy in the parallel rays will be focussed on receiver at X.

Transmitter

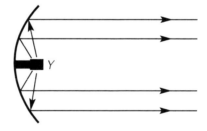

The energy from the transmitter at Y will travel out as a parallel beam, keeping the signal strong.

This principle is used in microwave transmitting and receiving dishes for mobile phones and TV.

Understand

Total Internal Reflection

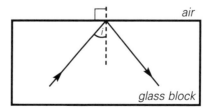

Total internal reflection occurs when **light in** a medium like glass or water **meets the boundary** with air and is **reflected back** into the medium.

This happens if the **angle of incidence, i is greater than the critical angle**.

If **angle i** is **less** than the **critical angle**, then the **ray passes out into the air (refraction)**.

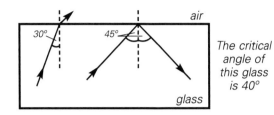

The critical
angle of
this glass
is 40°

Solve

The critical angle for the glass in these prisms is 42°. Find the angles marked i in each case then complete the diagrams to show that the light ray is refracted in diagram *a* and totally internally reflected in diagram *b*.

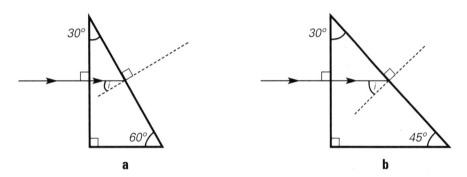

ANSWERS

a   i = 30°. This is less than the critical angle 42°, so the light ray is refracted.

b   i = 45°. This is greater than the critical angle 42°, so the light ray is totally internally reflected.

Understand

The principle of total internal reflection is used in optical fibres. Light pulses can travel through the fibres carrying information.

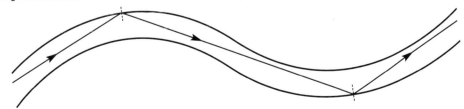

REFRACTION

Refraction occurs when light passes from one medium to another.

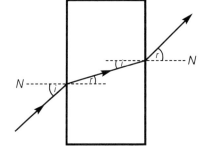

i - angle of incidence
r - angle of refraction
N - normal.

When **light goes from air to glass**, the ray bends **towards the normal**. This means that the angle of refraction is smaller than the angle of incidence.

When **light goes from glass into air**, the ray bends **away from the normal**. This means that the angle of refraction is **greater** than the angle of incidence.

LENSES

Converging lenses

Converging lenses bring parallel rays of light to a focus. When this happens, the **distance from lens to focus** is called the **focal length, f**.

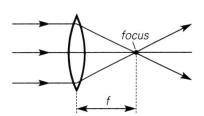 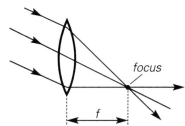

Diverging lenses

Diverging lenses cause parallel rays of light to diverge away from an imaginary (virtual) focal point.

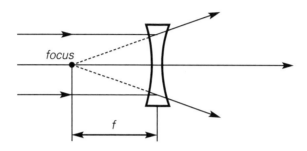

Memorise

The rules for drawing ray diagrams:

1   Draw the lens on a horizontal axis like this.

2   Mark the focal point on both sides of the lens.

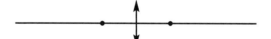

3   Draw the object as an upright arrow.

4   Any ray drawn parallel to the axis must pass through the focal point.

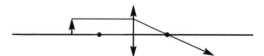

5   Any ray drawn through the centre of the lens will not change direction.

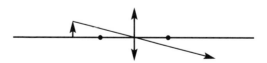

6   Where rays meet a real image will be formed.

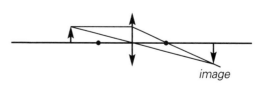

Memorise

An object placed more than two focal lengths from the lens will produce an image that is real inverted and diminished.

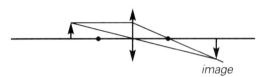

An object placed between one and two focal lengths from the lens will produce an image that is real inverted and magnified.

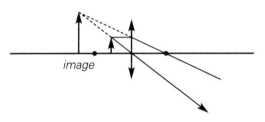

An object placed less than one focal length from the lens will produce an image that is virtual upright and magnified.

Solve

Complete these ray diagrams. State the nature of the image. Use a ruler to measure the position and size of each image. Which diagram shows the principle of a magnifying glass?

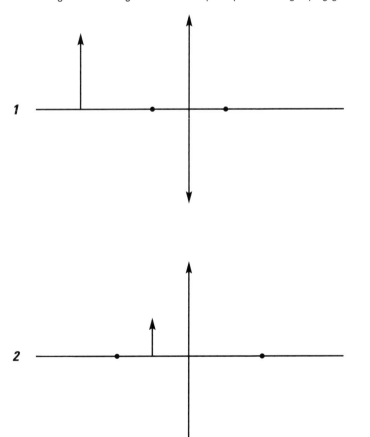

ANSWERS

1 The image is real and inverted, 1 cm high and 1.5 cm from the lens.

2 This diagram shows the principle of the magnifying glass. The image is virtual and upright, 2 cm high and 2 cm from the lens.

Memorise

$$\text{Power} = \frac{1}{\text{focal length in metres}} \qquad P = \frac{1}{f}$$

The **unit of lens power** is the **dioptre, D**.

Converging lenses have **positive** powers; **diverging lenses** have **negative** powers.

Solve

1   Find the power of **a** a 10 cm convex lens, **b** a 25 cm concave lens.

2   How far from a +2 D convex lens will parallel rays come to a focus?

ANSWERS

1   **a** $P = \dfrac{1}{f} = \dfrac{1}{0\cdot1} = +100$

    b $P = \dfrac{1}{f} = \dfrac{1}{0\cdot25} = -4D$

2   Parallel rays meet one focal length from the lens so we must find the focal length:

$P = \dfrac{1}{f}$ *so* $f = \dfrac{1}{p} = \dfrac{1}{2} = 0\cdot5\ m$

The rays meet 0·5 m or 50 cm from the lens.

THE EYE

Understand

People with **short sight** see objects a **short** distance away quite clearly. They have trouble seeing distant objects.

Rays of light focus at too **short** a distance to give a sharp image on the retina.

People with **long sight** see objects a **long** distance away quite clearly. They have trouble seeing objects close-up.

Rays of light would come to a focus at **too long a distance** to give a sharp image on the retina.

Short sight can be corrected using a **diverging lens**.

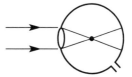

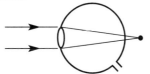

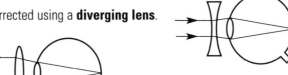

Long sight can be corrected using a **converging lens**.

*Chapter **4***

Radioactivity

PROPERTIES OF ALPHA, BETA AND GAMMA RADIATION

Describe

A simple model of the atom like the lithium atom shown in the example.

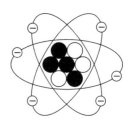

The atom consists of a central _____ containing _____ and _____ surrounded by _____ in _____.

(neutrons, orbits, protons, electrons, nucleus)

Memorise

Particle	Mass	Charge
proton	1	+1
neutron	1	0
electron	1/1840	-1

Electrons have such a small mass that we often write it as **zero**.

Understand

Alpha, **beta** and **gamma** radiations come from the **nuclei of certain atoms**.

An **alpha particle** consists of **two neutrons and two protons**. It is effectively a helium nucleus. $^{4}_{2}\alpha$

A **beta particle** is an **electron produced inside the nucleus** and immediately ejected from it at high speed. $^{0}_{-1}\beta$

A **gamma ray** is a **high frequency electromagnetic wave**. Sometimes, gamma rays are thought of as 'packets' of energy called photons. γ

Particle	Mass	Charge
α	4	+2
β	0	-1
γ	0	0

All three radiations carry energy.

When they pass through a material, their energy may be absorbed causing ionisation.

State

Ionisation occurs when an electron is knocked out of its orbit to produce a free electron and a positive ion.

Alpha particles produce very large amounts of ionisation, beta particles much less and gamma rays least of all.

Memorise

Absorption of the three radiations.

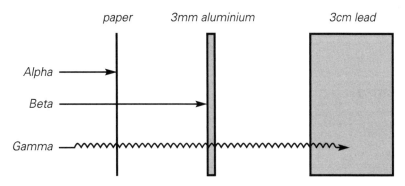

Describe

How a radiation detector works.

Ionisation detector: the Geiger-Muller tube

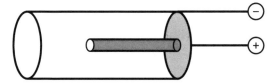

If an alpha, beta or gamma enters the tube, it causes_____. This produces a_____ electron which is accelerated away from the_____ charged outer casing towards the_____ charged central pin. The electron will crash into others on its way very quickly producing a large number of electrons. When this _____ of electrons hits the central pin a pulse of _____ will be produced. This will cause a single count on a counter.

(positively, negatively, avalanche, ionisation, free, electricity)

Scintillation detector: the gamma camera

If a gamma_____ hits certain types of crystal it can produce a tiny flash of_____ (a scintillation). The gamma camera has electronic circuits which can process all the small flashes into a _____.

(picture, ray, light)

State

Radiation can damage living cells. The ionisation it causes in the atoms of the cells can affect their chemistry, making changes to their DNA. In large doses, radiation may even kill the cells.

Describe

How radiation can be used to treat cancers

Cancer cells are more easily_____ by radiation than_____ cells. Sources of_____ radiation like cobalt 60 are contained in machines which_____ round the patient always targeting the_____ but _____the dose to healthy tissue.

(reducing, gamma, damaged, healthy, rotate, tumour)

The use of tracers

Patients can either swallow or be _____ with small quantities of radioactive gamma emitters. The amount of tracer_____ by a particular organ may indicate how healthy it is._____ placed outside the body can monitor the uptake of the radioactive material

(absorbed, detectors, injected)

Memorise

$$\text{Activity} = \frac{\text{number of decays}}{\text{time in seconds}} \qquad A = \frac{N}{t}$$

The **unit of activity** is the **becquerel, Bq**.

One becquerel equals **one decay per second**.

Solve

1 Find the activities of the sources X, Y and Z.

X emits 600 radiations every 10 seconds.

Y emits 360 000 particles in one hour.

Z emits $4\cdot2 \times 10^9$ particles every minute.

2 How many particles would be emitted by beta source of activity 20 kBq in **a** 10 s **b** one minute?

ANSWERS

1 X $\quad A = \dfrac{N}{t} = \dfrac{600}{10} = $ **60 Bq**

Y $\quad A = \dfrac{360\,000}{60 \times 60} = $ **100 Bq**

Z $\quad A = \dfrac{4 \cdot 2 \times 10^9}{60} = 7 \times 10^7 \text{ Bq}$ (or 70 MBq)

2 a $\quad A = \dfrac{N}{t} \; so \; N = At = 20 \times 10^3 \times 10$

$= $ **200 000 β particles**

b $\quad N = At = 20 \times 10^3 \times 60$

$= $ **1200 000 β particles**

DOSIMETRY

Memorise

$$\text{Absorbed dose} = \frac{\text{energy absorbed in joules}}{\text{mass of absorber in kilograms}} \qquad D = \frac{E}{m}$$

The **unit of absorbed dose** is the **gray, Gy**.

One gray equals one joule per kilogram.

Solve

1 Find the absorbed doses when:

   a A 70 kg scientist absorbs 0·21 joules of radiation.

   b A 60 g tumour absorbs 0·12 joules of radiation.

2 How much energy is absorbed when a 500 g piece of tissue receives in dose of 120 mGy ?

ANSWERS

1 a $\quad D = \dfrac{E}{m} = \dfrac{0 \cdot 21}{70} = $ **0·003 Gy** (or 3 mGy)

b $\quad D = \dfrac{0 \cdot 12}{0 \cdot 06} = $ **2 Gy**

2 $\quad D = \dfrac{E}{m} \; so \; E = Dm = 0 \cdot 120 \times 0 \cdot 5 = $ **0·06 J**

State

The biological risk from radiation depends on:

a   the **absorbed dose, D**

b   the **type of radiation** (alpha, beta, neutrons etc)

c   the **part of the body** exposed to the radiation

Understand

Each type of radiation is given a **quality factor Q** to show its biological effect. Here are some values:

Type of radiation	Q
α	20
β	1
Y	1
fast neutron	10

Memorise

Dose equivalent = absorbed dose in grays × quality factor

$$H = D \times Q$$

The **unit of dose equivalent** is the **sievert, Sv**.

Solve

Use the Q values in the table above to find the dose equivalents when:

1   **a**   A cancer patient receives 12 Gy of gamma radiation.

    b   A radiation worker receives 800 μGy of alpha radiation.

2   Find the total dose equivalent if a power station worker receives all of the following:

    300 μGy of beta radiation; 200 μGy of fast neutrons and 100 μGy of alpha particles.

ANSWERS

1 a $H = D \times Q = 12 \times 1 = \textbf{12 Sv}$.

   b $H = 800 \times 10^{-6} \times 20 = \textbf{0·016 Sv}$ (or 16 mSv)

2 $H = D \times Q = (300 \times 10^{-3} \times 1) + (200 \times 10^{-3} \times 10) + (800 \times 10^{-6} \times 20)$
   $= (300 \times 10^{-3}) + (2000 \times 10^{-3}) + (16 \times 10^{-3})$
   $= 2316 \times 10^{-3}$ Sv
   $= \textbf{2·316 Sv}$

Describe

The radioactivity in our environment is called _____radiation. About 90% of it is natural and 10% is_____. It comes from many sources including _____gas from rocks in the ground, cosmic radiation from space and even food and drink. On average, we are likely to receive a dose equivalent of about _____every year from background radiation although in some parts of the country this might be as high as 20 mSv per year. (At the other extreme, a dose equivalent of 10 Sv would probably be fatal to all humans within a few hours).

(radon, 2·5 mSv, man-made, background)

HALF-LIFE

State

The activity of a radioactive source decreases with time.

Half-life is the time taken for the **activity of a source to fall to half its original value**.

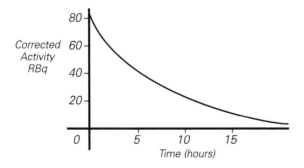

The half-life shown on this graph is five hours.

Solve

1 The activity of a radioactive source falls from 120 MBq to 7·5 MBq in 40 hours. Find its half-life.

2 Using the following data, draw a graph of corrected count rate in counts per minute against time and use the graph to find the half-life. (Background count rate equals 100 counts per minute).

Count rate	2600	1350	725	412	256	178
Time	0	3	6	9	12	15

ANSWERS

1 Activity (MBq) 120 → 60 → 30 → 15 → 7·5

The activity is halved four times in 40 hours
Four half-lives occur in 40 hours. The half-life is 10 hours.

2 **The half-life is 3 hours.**

Describe

The safety procedures when handling radioactive materials:

1 Don't touch radioactive materials at all if you are under ____ years of age.

2 Don't handle sources with bare hands; use_____.

3 Point sources away from the body and particularly away from the _____.

4 _____ hands after use.

(wash, tongs, 16, eyes)

State

Dose equivalent can be reduced by:

1 Shielding (lead, concrete etc)

2 Increasing the distance between yourself and the source.

3 Keeping the time of exposure as short as possible.

Memorise

The radiation hazard symbol.

NUCLEAR REACTORS

State

Some advantages of nuclear power stations.

1   They do not use up fossil fuels.

2   They do not release carbon dioxide or sulphur dioxide.

3   They use much less fuel than coal or oil fired stations. (One tonne of uranium is equivalent to about 25 000 tonnes of coal).

Some disadvantages of nuclear power stations.

1   They produce highly dangerous radioactive waste.

2   They are very expensive to decommission when their useful life is over.

Describe

Nuclear fission

When a _____ neutron collides with a uranium _____nucleus, the nucleus breaks up into two smaller parts called_____ . More neutrons are also released. Each time a nucleus breaks up_____ is released. If the neutrons strike further uranium nuclei then a_____ can occur.

(chain reaction, energy, 235, fission fragments, slow)

The nuclear reactor

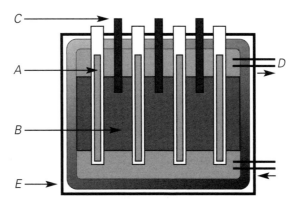

A   The fuel rods contain small quantities of uranium 235. The uranium 235 atoms produce energy when_____ occurs.

B   The moderator is made of_____. It slows down the neutrons.

C   The_____ control rods absorb _____ to maintain a steady, controlled amount of fission.

D   Coolant removes _____ from the reactor and transfers it to the boilers to produce_____.

E   The containment vessel is made of steel and concrete to withstand the heat and_____ and to protect workers from radiation.

(steam, boron, fission, pressure, neutrons, heat, graphite)

ABBREVIATIONS AND UNITS

Symbol	Meaning	Units	Abbreviated units
v	average (or final) speed	metres per second	m/s
s	distance	metres	m
t	time	seconds	s
a	acceleration	metres per second	m/s^2
u	initial speed	metres per second	m/s
w	weight	newtons	N
m	mass	kilograms	kg
g	gravitational field strength	newtons per kilogram	N/kg
F	force	newtons	N
p	momentum	kilogram metres per second	kg m/s
E_w	work done	joules	J
E	energy	joules	J
P	power	watts	W
E_p	gravitational potential energy	joules	J
E_k	kinetic energy	joules	J
E_h	heat transferred	joules	J
c	specific heat capacity	joules per kilogram degree Celsius	J/kg°C
ΔT	change in temperature	degrees Celsius	°C
l	latent heat capacity	joules per kilogram	J/kg
Q	charge	coulombs	C
I	current	amperes (amps)	A
V	voltage (potential difference)	volts	V
R	resistance	ohms	W
l	wavelength	metres	m
f	frequency	hertz	Hz
P	lens power	dioptres	D
f	lens focal length	metres	m
N	number of decays	*no units*	
A	activity	becquerels	Bq
H	does equivalent	sieverts	Sv
D	absorbed dose	grays or joules per kilogram	Gy or J/kg
Q	quality factor	*no units*	

EQUATION PRACTICE

It is useful to be able to change the subject of a formula when solving physics problems. The table below provides you with practice. Take a look at the first example. It is already worked out. Try the others:

	Equation	Rearrange to find	Working and answers	
1	$v = \dfrac{s}{t}$	t	$v = \dfrac{s}{t}$ *so* $s = vt$ *so*	$t = \dfrac{s}{v}$
2	$a = \dfrac{v-u}{t}$	t		$t =$
3	$a = \dfrac{v-u}{t}$	v		$v =$
4	$F = ma$	m		$m =$
5	$E_p = mgh$	h		$h =$
6	$E_k = \frac{1}{2}mv^2$	v		$v =$
7	$I = \dfrac{Q}{t}$	Q		$Q =$
8	$P = \dfrac{V^2}{R}$	V		$V =$
9	$P = I^2R$	I		$I =$
10	$\dfrac{V_1}{V_2} = \dfrac{R_1}{R_2}$	V_1		$V_1 =$
11	$\dfrac{N_s}{N_p} = \dfrac{V_s}{V_p}$	V_p		$V_p =$
12	$\dfrac{N_s}{N_p} = \dfrac{I_p}{I_s}$	I_p		$I_p =$
13	$Gain = \dfrac{V_{out}}{V_{in}}$	V_{out}		$V_{out} =$
14	$P = \dfrac{1}{f}$	f		$f =$
15	$A = \dfrac{N}{t}$	N		$N =$
16	$D = \dfrac{E}{m}$	E		$E =$

PREFIXES AND SCIENTIFIC NOTATION

μ	micro	10^{-6}
m	milli	10^{-3}
k	kilo	10^{3}
M	Mega	10^{6}
G	Giga	10^{9}

The table below provides you with practice in calculations with scientific notation. Why not put your calculator away, look at the worked examples, then try answering the others in the spaces provided.

1	365×10^{5}	3.65×10^{7}
2	0.09×10^{7}	9×10^{5}
3	$10^{4} \times 10^{7}$	10^{11}
4	$10^{9} \div 10^{12}$	10^{-3}
5	$10^{5} \times 10^{2}$	
6	$10^{-4} \times 10^{2}$	
7	$10^{-6} \times 10^{-11}$	
8	$10^{6} \div 10^{2}$	
9	$10^{6} \div 10^{9}$	
10	$10^{-4} \div 10^{-9}$	
11	$2 \times 10^{7} \times 3 \times 10^{5}$	
12	$4 \times 10^{-10} \times 2 \times 10^{4}$	
13	$5 \times 10^{11} \times 3 \times 10^{4}$	
14	$6 \times 10^{-7} \div 2 \times 10^{5}$	
15	$1.6 \times 10^{-19} \div 4 \times 10^{3}$	

ANSWERS

Equation practice

1	$t = \dfrac{s}{v}$	**9**	$I = \sqrt{\dfrac{P}{R}}$
2	$t = \dfrac{v - u}{a}$	**10**	$V_1 = \dfrac{R_1\, V_2}{R_2}$
3	$v = u + at$	**11**	$V_p = \dfrac{N_s\, I_s}{N_p}$
4	$m = \dfrac{f}{a}$	**12**	$I_p = \dfrac{N_s\, I_s}{N_p}$
5	$h = \dfrac{E_p}{Mg}$	**13**	$V_{out} = \text{Gain} \times V_{in}$
6	$v = \sqrt{\dfrac{2E_k}{m}}$	**14**	$f = \dfrac{I}{P}$
7	$Q = It$	**15**	$N = At$
8	$V = \sqrt{PR}$	**16**	$E = Dm$

Scientific notation

1 3.65×10^7

2 9×10^5

3 10^{11}

4 10^{-3}

5 10^7

6 10^{-2}

7 10^{-17}

8 10^4

9 10^{-3}

10 10^5

11 6×10^{12}

12 8×10^{-6}

13 $15 \times 10^{15} = 1.5 \times 10^{16}$

14 3×10^{-12}

15 $0.4 \times 10^{-22} = 4 \times 10^{-23}$

Index